SAVE ENERGY: SAVE MONEY!

I0170545

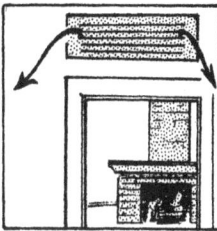

WILDSIDE PRESS

INTRODUCTION

Everybody seems to be worried about energy nowadays. There are some good reasons for this. The price of home heating fuels has almost doubled in the past two years and may yet go higher. For many people, this creates problems. Comfort and warmth are very important in the winter. Health depends on these -- particularly the health of older people, small children, and those who are sick.

Higher heating costs have come at a time when other prices have gone up as well. Inflation has driven food prices up. Even the prices of heat-saving materials such as insulation, caulking, and weatherstripping are going up. Many people can benefit from weatherizing their homes, and the time to do it is NOW! Most materials are available at your local hardware store.

In most cases, large savings can be made once people become aware of the problems and some low-cost ways of solving them. This booklet looks at the most common and most severe problems. <u>Directions are given for ways to save money which you can apply yourself</u>. These cost little and will help right away. Whether you own your home, rent a house, or live in an apartment or trailer, these simple ways to save energy and money can help you lead a more comfortable life.

You can start by helping yourself. A Bibliography has been added at the end for those who want to do further reading. Then help your neighbors by telling them what you have done. Those who cannot afford weatherization materials, or are disabled and cannot do the work themselves, may be eligible for assistance from their local Community Action Agency. Contact it for more information.

PLAY THIS GAME !

START HERE: Are your heating bills higher than you can afford?

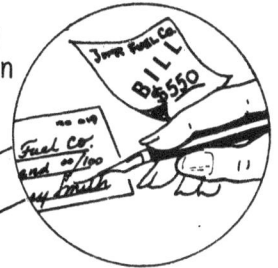

Can you see heat rippling up from your house on a cold day? This means that heat is escaping through the roof.

Does your house always seem cold and hard to heat on a windy winter day?

Even if you get a lot of heat from your heater, are some parts of the house very cold anyway?

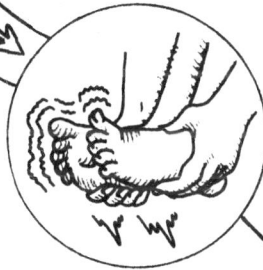

Are your floors cold, and are your feet cold in the winter?

Does there seem to be a draft of cold air moving down the stairs? Check with a candle. If so, your attic needs insulation.

Do most of your windows fog up with water or frost in the winter?

IF YOU ANSWERED "YES" TO TWO OR MORE OF THESE, YOU'VE GOT ENERGY TROUBLES!

KEEPING WARM

The main reason for high heat bills is that cold air seeps into the house through cracks and holes. Whether the cracks and holes are large or small, they all rob you of heat.

Especially on windy days, these leaks make the house very hard to heat. Cold drafts, particularly along the floor, will make you uncomfortable. If you have to pay for heat, leaks can cost you up to 20% of your heat bill in an older home! So let's button up for winter -- keep warm and comfortable -- and save lots of money as well.

SEALING HOLES

These can be anywhere in your house or apartment. Look at windows, walls, ceilings and roofs. If you find holes, seal them up.

Check for broken or cracked glass. If the glass is broken, try to put a new piece in with putty. Be careful to remove the old glass with gloves or a cloth, so you don't cut yourself.

REMOVE BROKEN GLASS WITH GLOVES, CLOTH, OR PLIERS

ADD PUTTY

USE GLAZIER POINTS TO HOLD GLASS, THEN PUTTY

Sometimes it's easier to patch the glass if the hole is small or the glass is just cracked. Freezer tape or duct tape criss-crossed over the hole or taped along the crack works just fine.

Look for holes in your roof and walls. One sure sign is the presence of water-marks on ceilings and walls. You can repair minor leaks and small holes with a "sealer compound," available in most hardware stores. You should patch larger roof holes with tar-paper, asphalt, or shingles. Outside walls can be patched with tar-paper spread over sealer, or mortar can be used on concrete and brick walls. Call your local Community Action Agency if you need help with this work.

On the inside of your house, look for holes which let air go up through the ceiling. Make sure that the attic door closes tightly. Are there openings around pipes or flues which go up through the ceiling? All of these let heat escape.

You can use newspaper or cloth to seal around the edges of attic doors. Cloth is better for sealing around pipes. Stuff it into the hole. If these are holes where the chimney or flue goes through the ceiling or wall, do NOT use cloth. Cloth could catch fire. These holes must be patched with an asbestos sleeve.

SEALING CRACKS

Many small cracks around doors and windows allow heat to escape and let cold air into the house. If you put all of them together, they let as much heat escape as a large open window would! So it's important to seal these up. Let's look at ways we can close up small cracks and holes.

CAULKING -- Check around window frames, door frames and chimneys. The easiest way to fill cracks here is with caulking. First clean away any grease, dirt, or old caulking. Do this work early in the season or on a warm day, as the temperature must be above 45°.

The best caulking is the "acrylic-latex" type. It is more expensive than regular caulk, but the seal will last five years or more. Small cracks can be filled with "cord caulking."

CAULKING GUN

Usually caulking applied with a "caulking gun" works best on larger cracks. It makes the whole job very easy. Just apply the caulk by pushing the gun along in the direction of the crack. This makes the best seal.

Large or deep holes can first be filled with "oakum." This is a rope-like material which you can buy at hardware stores.

WEATHERSTRIPPING -- Check for air leaks on a windy day. Place your hand near doors and windows. Can you feel air coming in where the window or door meets the frame? Now stand on a chair. Can you see light through the crack above the door? If your answer is "yes" in either case, you need weatherstripping.

On windows, most of the air will come through the cracks at the top, bottom, and center where the window meets the casing. You can place folded cloth or newspaper in these areas. Then close and lock the window to make a tight seal. Check again with your hand. If you still feel cold air at the sides of the window, buy some "tubular gasket" weatherstripping at the hardware store and tack it along the edge where the window slides up and down. This weatherstripping is cheap and really stops air leaks.

SEAL CRACKS WITH CLOTH

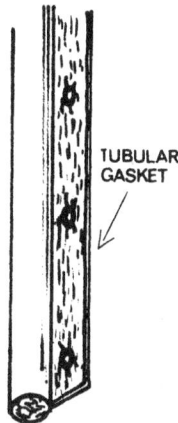

WEATHERSTRIP YOUR WINDOW

Also, check the casing all around the window. If there are open seams between it and the wall, tighten the casing with predrilled screws or nails and caulk if you need to.

On doors, add molding at the top and sides if it isn't already there. Once this is in place, tack some tubular gasket weatherstripping to the molding. The door now makes a good seal against the molding when it is closed. The crack at the floor can be sealed with a throw rug, though you must replace it each time. You can also use a strip of "reinforced felt hair" here. And always remember to open and close the door quickly in winter. Let as little heat out as possible when you use the door!

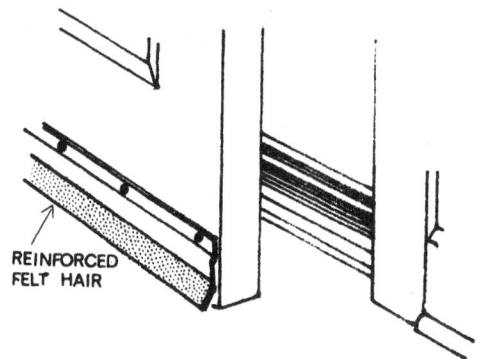

7

SEALING FOUNDATIONS AND FLOORS

Many older houses develop air leaks through the floor. One problem is that heat escapes. Another problem is that cold air can blow in from below. Houses on poles and those with stone foundations let the wind blow underneath the floor. Here are some things you can do to keep Old Man Winter out:

Cover your floors with old rugs. If you don't have enough, these can be bought cheaply at garage sales and auctions. Before you lay the rug down, cover the floor with several layers of newspaper and then tack down the rug.

If the foundation around your house or trailer is open so the wind can get in, enclose it with wooden "skirts." This can also be done by "banking." Fill sacks with dirt and stuff these around the edge of the house to fill the space between the sill and the ground.

Another way is to nail tar paper to the sill and drape it to the ground. Hold it in place with rocks, bricks, or dirt.

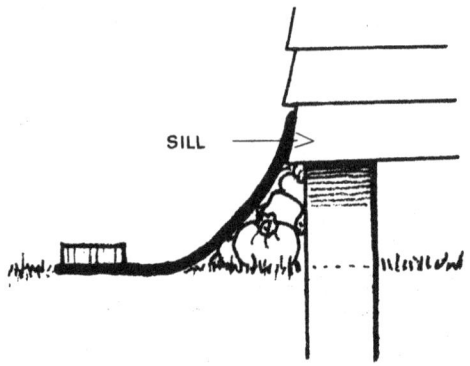

FRONT VIEW   (IT'S HELPFUL TO PLACE A FEW SACKS   SIDE VIEW
              OF DIRT BENEATH ENDS OF TAR PAPER
                     TO BLOCK THE WIND)

Whichever way you do it, make good use of the snows when they come. Shovel snow over the sacks or tar paper to fill in any air spaces. The snow also helps keep heat in.

8

STORM WINDOWS

The next step in cold climates is to put up storm windows. These help with air leaks and also slow down the loss of heat through the windows.

If you can, buy some old storm windows at a garage sale. You might get enough for the whole house for about $10. Be sure to measure your windows first, so that the storm windows you buy will fit tightly.

If you can't buy used storm windows, you can purchase low-cost acrylic or aluminum framed storm windows. Check with your local Community Action Agency, as they may be installing these as a part of their weatherization program. A temporary solution is to use plastic film. The plastic lasts only one winter, but it is cheap. Get a roll of plastic wide enough to stretch across your windows. Be sure to use a heavy plastic. A thickness of 6 to 10 "mil" is good. A thin plastic, only 2 to 4 mils thick, may blow down before the winter is over.

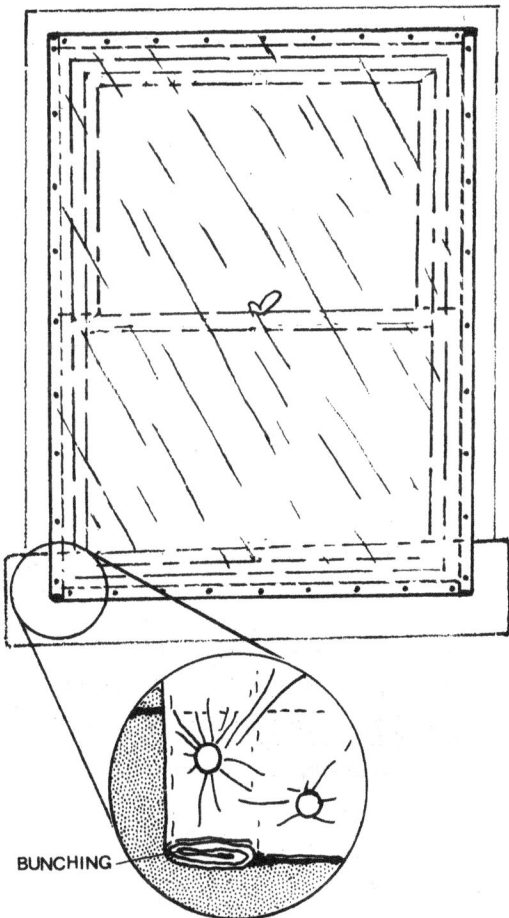

BUNCHING

For hard-to-reach areas, put the plastic inside the house. This works well in apartments and on upper floors.

Make sure you weatherstrip the window before you add the storm window. Weatherstripping stops moisture in the house from getting between the glass and the storm window, and frosting them up.

The storm window must fit tightly. For frame storm windows, add some folded paper or weatherstripping to the frame. Then fasten the storm windows in place to make them very tight.

For plastic storm windows, fold the plastic several times around a thin piece of cardboard or wood at the edge, and tack or staple through the "bunching." This bunching helps give a good air-tight seal all around the window.

FOLD THE PLASTIC FOR A TIGHT SEAL

TURN DOWN THE HEAT

After you "button up" the house, turn down the heat. You will need less now. Try to keep the heat as low as you can all winter. The best way to do this is to get used to a lower temperature slowly, as winter comes on. Most people do fine at 65° or less, but be careful if there are small babies crawling on cold floors, or older folks who might catch colds. Go slowly, a little at a time. Let the whole family get used to lower temperatures.

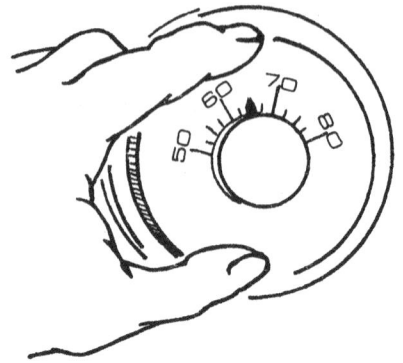

Living with less heat can be easy. The best way is to dress up yourself, just as you buttoned up the house.

CLOTHES -- Make sure clothes are loose fitting. This traps body heat and helps keep you warm.

Use wool and cotton clothing. These are much warmer than man-made fabrics like nylon and rayon.

Wear several layers of light clothing. This traps warm air and makes you feel more comfortable than one heavy layer would. Long underwear can be a big help. Wear sweaters for an extra layer of protection.

When you sit and work or watch t.v., use a quilt or blanket over your legs.

BEDCOVERS -- Use quilts and comforters on beds to keep you warm at night. Just as with clothes, several light blankets trap warm air and keep you more comfortable than one heavy cover. Socks and long johns are great. Also, don't forget that grandpa used a nightcap to keep his head warm.

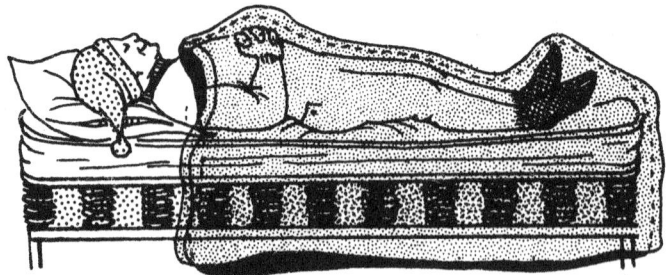

Electric blankets are a good investment. They use only small amounts of electricity. You can turn the house heat down at night, to 60° or below, and still stay warm. Use a sheet, the electric blanket, and heavy covers on top to keep the heat next to your body.

Pin blankets and covers in place so they don't slip off. This is especially important on children's beds. But be careful with electric blankets. Pin only the corners, where there is no wiring, or you may get a shock.

If you sleep on a mattress on the floor, make a platform with scrap wood. It should be at least two feet high. The platform will lift you up above the cold air next to the floor. It's much warmer on the platform, and the kids can avoid the drafty floor by using it as a play area during the day.

DRYNESS

The winter is a dry time in most houses. You feel this when your nose and throat dry out and become sore. You can catch colds from dryness. Sealing air leaks will help by keeping moist air from the kitchen and bath- room in the house. Once you have done this, try a few other things, if they are necessary.

Place metal pans or cans of water on radiators or heaters. (Never on electric heaters!) The water puts moisture into the air. Also, keep house plants to freshen and add moisture to the air. Dry your clothes at home near a radiator or stove. On sunny days, dry clothes near a south window.

Remember: Buttoning up the house, dressing for winter, and turning down the heat has saved you a lot of money. It can save about $20 or more out of each $100 of your heating bills!

If you spend a little money for weatherstripping and caulking, you will get all this back, in less than one winter, from the money you save on fuel bills.

The next big savings come from insulating. If your attic doesn't have insulation, putting it in could save as much as another $15 out of each $100 you usually spend for heat. (See page 29.) Insulation costs money to install, though -- perhaps $150 for the attic of a small house. If you rent an apartment or home there are still things you can do to save money and make your home more comfortable. Also, if you do own your home but don't have the money to fully insulate it right now, use the information in the rest of this booklet. Then you can buy some insulation next summer with the money you save this winter. So let's look next at some low-cost ways to make heat and other energy savings.

GETTING HEAT WHERE YOU NEED IT

Once you have sealed the air leaks in your home, use the heat you have to give yourself the greatest comfort. Learn how to get the heat where you need it when you need it.

HEAT ONLY THE PARTS OF THE HOUSE YOU USE!

WHEN YOU'RE AWAY -- When you leave the house, even for a few hours, turn the heat down. If you're away for several days, lower your heat to about 50°. When you return, it will take awhile to warm up the house again. Try to schedule unpacking or housework for these times. Don't turn the heat up past its usual setting. That wastes some of the fuel and money you saved while away.

WHEN YOU'RE HOME -- Heat only the rooms you are using. The heat in bedrooms can be turned down, or turned off altogether in warm climates. Use lots of bedcovers at night, along with the heat rising up from downstairs, to keep yourself warm while sleeping. During the day, close off doors to bedrooms and heat only those rooms you are using. If you don't have bedroom doors, a heavy blanket tacked at the top and one side of the opening will help.

You can separate the upstairs from downstairs in a better way by putting vents in the upstairs floor. Different sizes are available at hardware stores. Cut holes in the upstairs floor and downstairs ceiling. Place the vents in. Close them each morning with a throw rug to keep heat from going up to the bedrooms. Remove the rug each night to allow heat to rise into the upstairs part of the house.

SAW VENT-SIZE HOLE BETWEEN STUDS

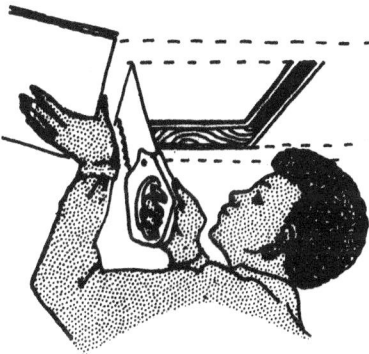

PUT VENT IN PLACE

ADJUST AIR FLOW WITH THROW RUG

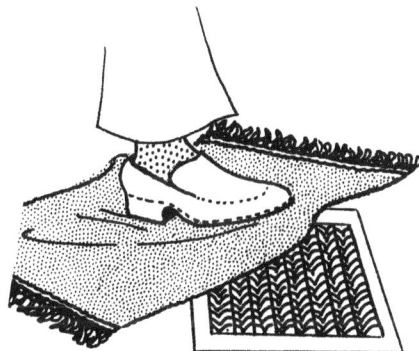

Less heat is needed in work areas like the kitchen, because the physical activity will keep you warm. So turn down the heat a bit in these rooms, and save.

CLOSE OFF DOORS, WINDOWS & ROOMS
WHICH ARE NOT USED!

Take a look at your house or apartment. Very little light comes through the windows on the north and west sides in winter. Yet the cold winds blow from this direction and rob you of heat. If there is more than one door, close the ones on the north and west sides for the winter. The same should be done with extra windows and extra rooms on these sides of the house.

LESS HEAT IS NEEDED
IN WORK AREAS

To close off doors to the out-side -- first fill any cracks around the edge with folded newspapers or cloth. Then tack a plastic storm window onto the door frame to ensure that all air leaks are stopped.

Windows lose heat much more quickly than walls. Windows you don't need should be covered with plywood screwed to the window frame. Add newspapers, cloth, or weatherstripping between the wood and the window to get an airtight fit.

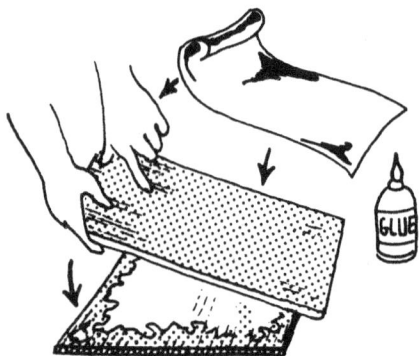

Another way to cover windows is to make one large cardboard piece by gluing together smaller ones. A white glue works fine. Four layers of cardboard helps to insulate as well, especially if you put wadded-up newspaper between the cover and the window before tacking the cover onto the window frame. You can glue some aluminum foil to the cover on the part that faces inside. This will reflect the heat away from the window area and into the room. A white sheet draped in front of the cardboard makes an attractive cover. Keep a half-inch air space between the aluminum and the sheet, for insulation. Now, turn down the heat if there is a radiator, register, or baseboard heater below this window. The plywood or cardboard cover will greatly reduce the cold air that usually builds up near the window.

14

To close off extra rooms -- first, make certain that a thermostat which controls the heat for other rooms is not located in the room you intend to close off. Then cover the windows in the extra room with plywood or cardboard. Now turn the heat off in that room. Shut off valves on radiators till they are tight. If you have hot air registers, close registers and block the heat further with a towel or throw rug. Close the valves on baseboard radiators. If there are no valves, block the heat flow by closing the metal vanes on top of the baseboard unit.

Now close the door, using newspaper or cloth to make a tight seal. It's a good idea to check these rooms at first. Make sure the temperature stays above freezing. Also, moist house air will sometimes get into the cold room. This wetness can damage paint, wallpaper and furniture. Open the door after a few cold days and take a look. If this isn't a problem, you can leave the door closed all winter. But if it is, leave the door open a few hours each day.

Rooms without doors can be shut off, too. Add a door if you can. Another way is to tack a blanket around the frame of the opening, or use cardboard or a plastic sheet as you would on a window.

GET THE HEAT TO COLD AREAS OF THE HOUSE!

Some essential areas of your house or apartment may be cold spots. This can be true even when other rooms near the heater are very warm. Don't turn up the heat. This doesn't help much and it overheats some rooms. Try these other things first:

Check to make sure that nothing is blocking the heat in the cold room. Are radiator valves open and is the area around the radiators clear of furniture which might trap heat? If you have a hot air system, be sure the registers are fully open and not blocked by furniture. Check that the furnace filters are clean. If you still have a problem, reduce the flow of heat in the warmest rooms closest to the furnace. This should help increase the flow of hot air through the registers in the cold room.

Move furniture away from the outside wall or the windows in the cold room. Except by sunny south-facing windows, it should be more comfortable to sit or work near an inside wall, where it's warmer.

15

If the room still seems cold, use it as a storeroom. You can turn a cold spot to good advantage this way.

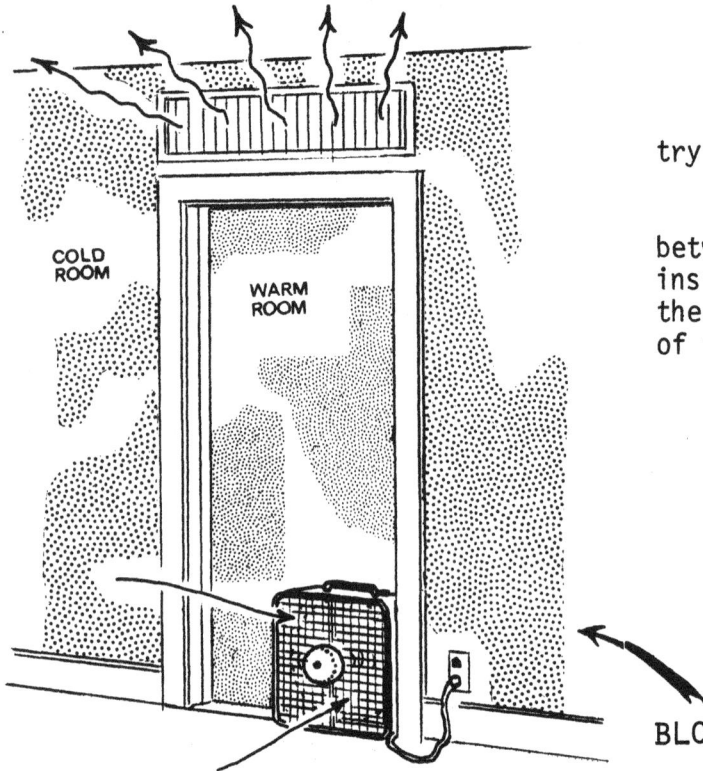

If you _need_ the area for work, try one of these ways of heating it:

WALL GRILLE -- Cut a hole between the door and ceiling and install a grille or vent. This lets the hot air trapped on the ceiling of the warm room into the cold one.

BLOW COLD AIR _INTO_ THE WARM ROOM

FAN -- If you still have a problem, use a fan to move air from room to room. Place the fan on the floor in the doorway between the hot and cold room. Blow air from the cold room _into_ the warm room. The cold, heavy air is easier to move and will force warm air through the door and heat the cold room quickly. Be _very careful_ when using the fan. Small children may want to play with it.

MOVE HEAT AWAY FROM THE HEATER!

Try to get more heat out into the room where it can be used. Don't let it get trapped behind or above radiators and heaters.

Keep heat away from the walls behind radiators and wood stoves. This can be done with reflectors made of thin sheet metal which direct the heat out into the room. Aluminum foil taped to the wall or glued to a piece of plasterboard will do fine, too.

Direct the heat away from the ceiling and into the room. Radiators are usually placed by windows so that they heat the cold air around them. This heat rises to the ceiling where it is of little use. Make a reflector of thin sheet metal and aluminum foil. The foil should be taped to the wall behind the radiator. The sheet metal should be curved, as in the picture, to push warm air out into the room.

WALL REFLECTOR OF PLASTERBOARD AND ALUMINUM FOIL

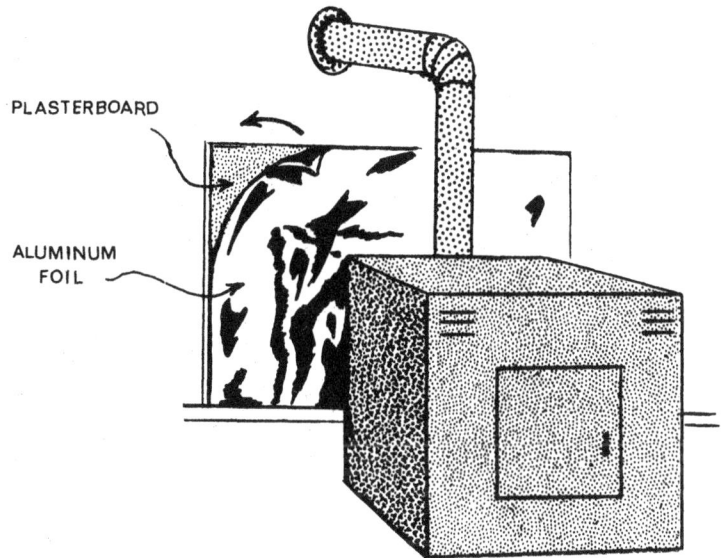

PLASTERBOARD

ALUMINUM FOIL

MAKE RADIATOR REFLECTOR TO GET MORE HEAT

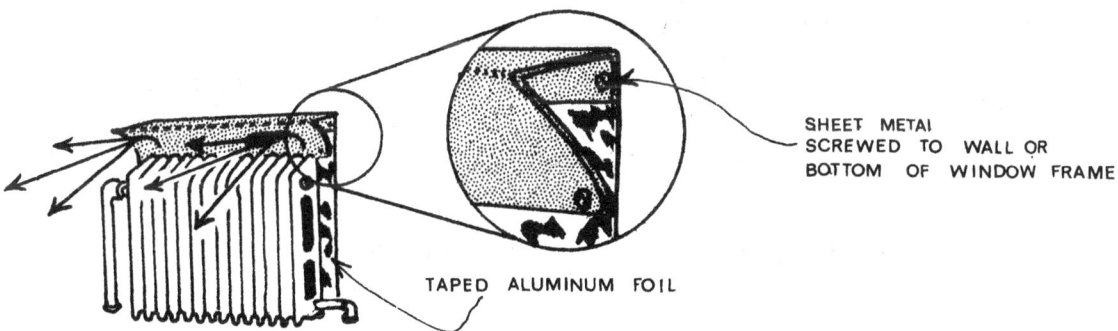

SHEET METAL SCREWED TO WALL OR BOTTOM OF WINDOW FRAME

TAPED ALUMINUM FOIL

Remember: Turn down the heat when you're not home. Close off doors, windows, and rooms which you don't use. If some essential areas of the house are cold, move heat from warm rooms and away from radiators and stoves to heat the cold areas. Doing these things will make your house more comfortable. It will also save you some more money, maybe another $10 of each $100 you usually spend for heat.

17

USING THE SUN

Button up the house and spread the heat where you need it. Then, sunny winter days will hold a warm surprise for you. Make every use you can of the sun's heat in winter. It's free and the warm light is comfortable and relaxing.

LET THE SUN SHINE IN

Now that you have reduced the cold drafts near windows by weatherstripping, use the sun whenever there is a clear day in winter. The windows on the south side get the most light. Windows on the east and west sides also get some. From the east, sunlight comes more in the morning. On the west side, you will get afternoon sun. Open curtains and blinds and let this sun into the house. Make sure that there is nothing near these windows which will block light from coming into the room.

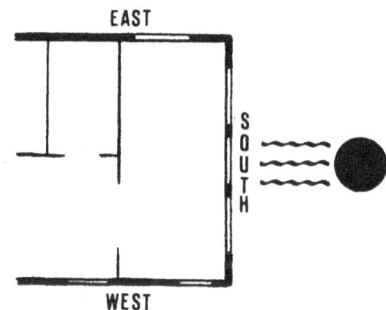

If there are screens on these sunny windows, remove them for the winter so they don't block the sun. Try to work and relax in the sunny rooms during the day. They are the warmest rooms in the house and the extra heat is free.

USE CURTAINS

To make the best use of sunny windows, you must close them off when the sun is not shining. Pull down light-colored shades or close venetian blinds so the bottom rests on the windowsill. Better yet -- put up curtains, and then close them off at night. This is very important, particularly in cold northern climates. Windows on all sides of your home lose heat to the out-of-doors very quickly. They can rob you of heat two or three times faster than even an uninsulated wall of the same size! So when the sun goes down, close off windows!

Another time to curtain off a window is during the summer. Also, put your screens back on in summer. They block some of the light and keep out insects. If you have awnings, use them. They shade the window and keep the house cooler.

MAKE A GOOD WINDOW CURTAIN

Good window curtains will help you save money, and help keep your home warm and comfortable, too. They reflect heat back into the house and keep warm air away from the cold windows. To do this, the curtain <u>must</u> be tight at the top, bottom, and <u>sides</u>! Otherwise, the warm air at the ceiling will move down the walls and windows as it cools. So button up the curtain on all sides and you will have a great heat-saver.

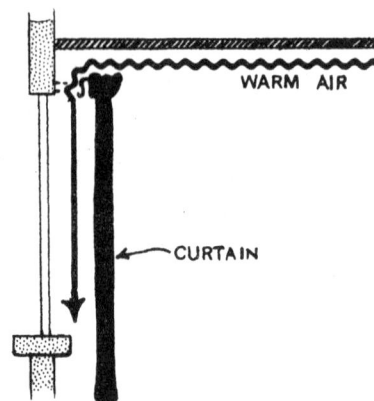

AIR IN CONTACT WITH
WINDOW LOSES HEAT

The curtain should also be light-colored, to reflect back more light and heat into the room. The curtain material should be thick enough so air can't go through it easily. A bed-sheet would be too lightweight and would let air pass through, but an old wool blanket would be fine. Your curtain will insulate better and save even more heat if you sew a curtain liner to the blanket. Or, if you have an old quilted comforter, use that.

You can make your own curtain, and cheaply too. Make it so that it seals air out of the top, sides, and bottom when it is closed. Here is the way to do it. To make the top "cap" for the curtain, attach a piece of material to the window frame several inches above the curtain rod area. It should drape over the top of the curtain all the way around. Cut the material so there are no "ruffles". The cap should lie flat on top of the curtain to make a good air seal. Also, when you tack the cap to the window frame, make sure it seals well.

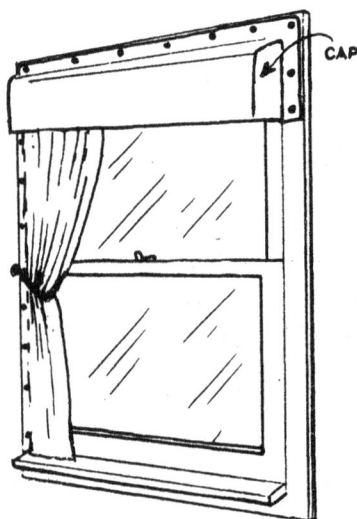

CAP YOUR WINDOW TO
SAVE HEAT

Tack one side of the curtain to the window frame to make it airtight. On sunny winter days the curtain can be pulled and tied to this side. Tie it above center, as this clears more of the window to let light in. The curtain must slide along the rod to do this, so make sure it can move. Little rope loops will be fine. Also, make sure the curtain rod is smooth and round so the curtain can slide easily.

When you pull the curtain across the window at the end of the day, there will still be a space on the bottom and on the other side where air can get next to the window and cool down. If you want to keep in the most warmth with your curtain, you have to make a tight seal here too. This can be done simply, with wood screws and sturdy lumber yard "furring strips." The furring strips hold the curtain against the window frame, using two wood blocks as in the picture. Adjust the blocks so that the curtain is just tight enough against the window frame to give a good air seal. You have only to pull the furring strips to one side to move the curtain.

This kind of curtain will really reduce your heating bill. In summer, use it to trap heat at the window and help keep your house cool. Just open the window a bit at the top and bottom, and the closed curtain will block the sun and keep the hot air next to the window where it can escape to the outdoors.

STRIPS SWING OUT TO RELEASE
CURTAIN
(WINDOW SHOWN WITHOUT CURTAIN CAP)

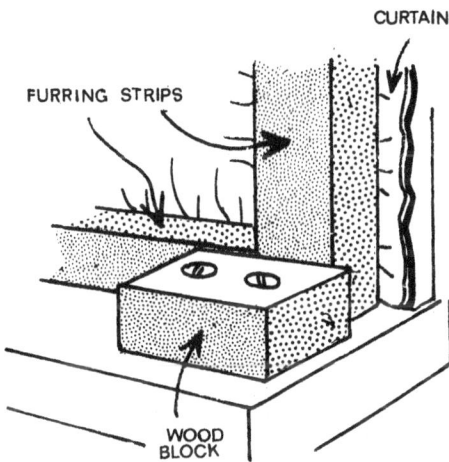

CURTAIN

FURRING STRIPS

WOOD
BLOCK

STRIPS SWING BACK TO
HOLD CURTAIN TIGHT

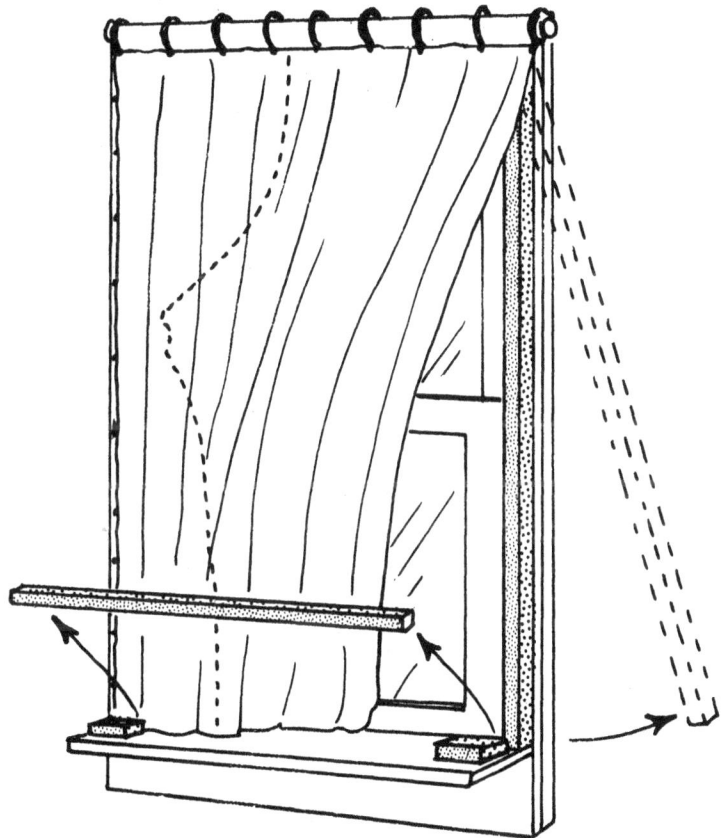

Remember: Use the winter sun when you can. Close off curtains at night and keep the heat in your house. Whatever you spend on materials to make good curtains, as described above, will come back to you in one or two winters as money saved on heating bills. In the summer you can also use curtains to help keep the sun out.

FURNACES, STOVES, & FIREPLACES

Now that your home is buttoned up, you have heat where you need it, and you know how to use the sun, your heating bills will be much lower. The next step in saving heat is to learn to get all the heat you can from the fuel you burn!

KEEP HEATERS CLEAN

CLEANING COAL FURNACES AND STOVES -- Clean these before winter starts and at least once more during the heating season. Work on a warm day. Put the fire out completely before you start.

Use a broom or brush to knock soot off the top and sides of the firebox.

Clean out the soot from the ashpan and the top of the grate.

If you have an oil or gas heater, have a competent repairman clean and adjust it before winter comes. This check-out will cost some money, but it's worth it. You will get more heat from less fuel.

If you use coal, you can do this job yourself and it will save you money. The first step is to clean the furnace or heating stove. Then clean the flue and the chimney. This cleaning will help remove soot and creosote which might cause a chimney fire. Remember, cleaning your heating system is a dirty job -- so dress for it.

FLUEPIPES -- While the fire is out, remove the fluepipes and clean them. Mark each section of pipe between the heater and the chimney with chalk or marker pen. This makes it easier to put back together.

Tap each section of pipe to loosen the soot.

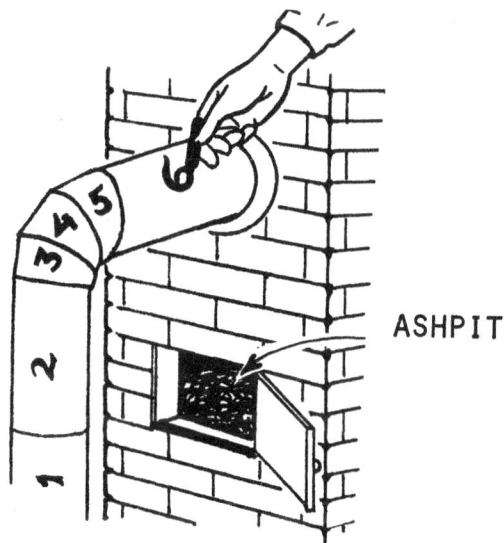

ASHPIT

Brush out the soot. An old broom will do a good job. Cut part of the binding so some of the straws stick out. Twist the broom and push it back and forth in the pipe.

CHIMNEYS -- Next, clean the chimney. Once each season should be enough. Cover the hole where the fluepipe fits into the chimney. This will stop soot from falling out into the house while you clean.

Get on the roof and carefully move to the chimney. Tie a rope around the chimney and fasten the other end around your waist. Tie it securely so you will not slip or fall.

You can clean the chimney in several different ways. Fill an old burlap sack with bricks so that it loosely fits into the chimney. Attach a rope and lower it down the chimney a few feet. Now pull it up and down. Once this area is clean, drop the sack a few more feet and clean again. Do this until you get the whole chimney clean.

If your chimney is capped, you can still clean it. Tie a tire chain to a piece of rope. Lower it down the chimney and rattle it around to break the soot loose.

CLEANING A CAPPED CHIMNEY

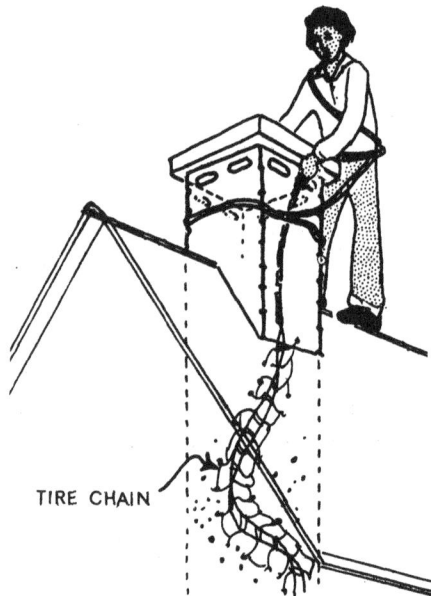

TIRE CHAIN

If you are cleaning a chimney with a large opening, a fine way is to scour it with a small birch tree, or the upper limbs of a larger birch tree. One person standing on the roof and another in front of the fireplace, can haul the tree back and forth on ropes. A birch tree has stout, spiky limbs, like a big wire brush.

Whichever way you use, be sure to clean the soot from the ashpit or fireplace, when you are done.

FINISHING THE JOB -- Put the sections of fluepipe back together and attach the flue to the chimney and heater. Use the numbers to get each section of pipe where it belongs.

Check the pipe joints for leaks. If the pieces don't fit together well, tap out dents with a hammer and wrap some asbestos tape around the joint, if necessary. You can buy this kind of tape at a hardware store.

KEEP YOUR CHIMNEY AND FLUE CLEAN -- During the heating season, burn a little block of "creosote and soot remover" as often as once each month in your furnace or fireplace. This will really help prevent the chance of a fire in your chimney or flue.

OTHER THINGS TO DO

AIR VALVE

If you have a hot water or forced air heating system, check these also:

Vacuum dust off your radiators. If the paint on the radiator is old and flaky, scrape it off with a wire brush. Dust and flaky paint are very bad. They block heat. Bleed hot water radiators to get air out of the system. Open the air valve until water just starts to come out. Then, close it tightly. Use a screwdriver, a coin, or a radiator key for this. On forced air systems, clean the filters once a month with a vacuum cleaner. Also, vacuum over the outlet to remove any dust or dirt.

CLEAN FILTERS MONTHLY

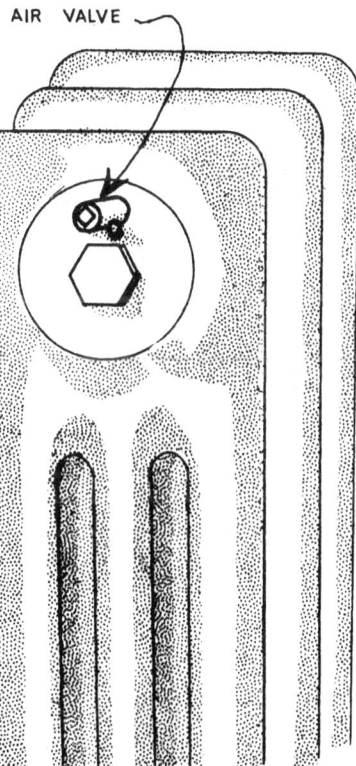

BLEED RADIATORS

Make sure that all heating pipes or hot air ducts have insulation on them where they run through the basement or cold rooms. Pipe insulation or wraparound insulation for these ducts can be bought at a hardware store. It's easy to put on and can really save money for you.

FIREPLACES AND HEATING STOVES

Both fireplaces and wood or coal stoves can be used to heat the whole house. Or you can use them for extra heat. When burning wood, you must keep your flue and chimney clean just as you must with a coal furnace or stove. This is very important, particularly if you burn wood which is not fully dry. When you have a choice between a fireplace and a stove, use the stove. Stoves give off more heat. If there are two chimneys in the house, use the one on the inside wall. It will draw better and give more heat, too. Burn coal only in a stove that was designed for its high heat.

Fireplaces use a lot of air to keep going. Fresh cold air will seep in around doors and windows to replace what goes out the flue. So a fireplace can even cool your house down! For the greatest comfort, close some doors and try to draw air in through empty rooms. Another solution is to build a fireplace air vent. This vent is explained in the "Do-it-yourself" section of this booklet. Don't build roaring fires! They waste fuel because the heat goes up the chimney before you can use it. Burn wood slowly.

Close the damper when the stove or fireplace is not in use. Leaving the damper fully open will waste a lot of heat. Be certain the fire is completely out before you close off the damper. If it isn't, leave the damper open a little so the flue gases can move up the chimney as the fire dies out. There are a couple of things you can do if your heater doesn't have a damper. For stoves, you can buy a special section of stovepipe with a damper in it at many hardware stores. With a fireplace, some people stuff newspaper a little way up the flue. Make sure to pull the paper down and use it to start the next fire, if you do this. Otherwise you might smoke up the house or cause a fire hazard. A safer and still low-cost solution is to make a sheet metal shield to fit over the fireplace opening when it is not in use. Buy some fasteners at the hardware store so that you can put up and take down the shield easily. Cut several small holes about 1" wide near the bottom of the shield. These will allow a fire to smolder behind the shield without smoking up the house after you go to bed. Cover the holes with bricks when the fireplace is not in use.

ADD A WOOD STOVE

Adding a small wood stove can be a low-cost way of getting some extra heat. Small flat-top "chunk stoves", also called "tin stoves" or "Dover stoves" can sometimes be purchased for as little as $25 in hardware stores. You can use the flat top to cook and heat water. If you stay away from building roaring fires in these small stoves, there should be no problems when you add an extra flue to your chimney. If the flue touches materials which

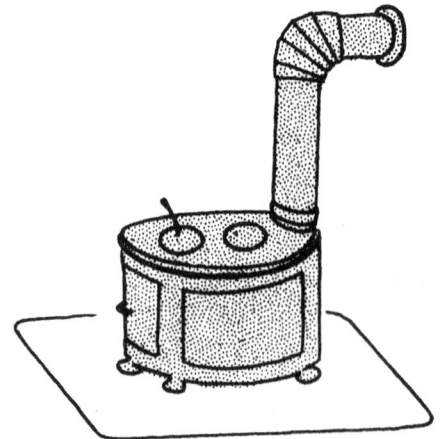

A CHUNK STOVE

could burn where it goes through your wall, use an asbestos sleeve for fire safety. Also, check your local codes. The flue cap or top of the chimney usually must be at least two feet above the roof ridge for safety. Finally, cover the rug or floor under your new stove with an asbestos-sheet metal stove reflector. This reflects heat up and protects the floor.

You can burn just about anything in "chunk stoves." You can burn paper, logs, small chunks of wood, or broken-up wooden boxes. Try to collect wood in the spring so that it has time to dry. "Resinous" woods, like pine or fir, should be allowed to dry for a year. Dry wood is easier to burn and gives off more heat. There is also less danger of a chimney fire. When you store wood, protect it from the rain. It will dry more quickly that way.

A nice way to make use of extra newspapers or magazines is to make paper logs. Roll up the paper in round log-like shapes. Tie these with string and let them soak in water until fully wet. Dry them out in the house. The moisture helps fight winter dryness. Once dry, paper logs burn almost like wood and keep a good fire going for quite some time.

If you use wood as your major source of heat, an efficient wood space heater would be a good investment. First of all, a chunk stove could burn out in a few years, while a good wood space heater is likely to last a lifetime. They are more expensive than chunk stoves. A good wood space heater costs at least $150 -- but it will have an efficiency of about 50% or more!

MAKING PAPER LOGS

Remember: Get all the heat you can from the fuel you burn! Keep your heating system clean and burn the fuel safely and slowly. This will improve the efficiency of your heater. You can save another $10 out of each $100 you usually spend for heat. You can save even more money if you add a small stove and use wood for part of your heat.

INSULATION & OTHER ENERGY NEEDS ▶

Help yourself! You can save heat and lots of money. You've buttoned up your house. This has saved you up to $20 out of each $100 you usually spent for heat. You have put the heat where you need it, and you now use the sun and have put good curtains over your windows, saving as much as another $15 out of each $100 you spent for heat. You also now get all the heat you can from your fuel, saving as much as $10 more out of each $100. Doing all these things has saved you a lot of money -- a total of perhaps $45 out of each $100 you used to spend for heat!

Now, use these savings to help you save even more money. Make sure that your attic is well insulated. You can save up to $15 or more out of each $100 this way.

If you can't afford to fully insulate your home all at once, start with a little insulation in the attic. Three to four inches of insulation is a good start. This much insulation material would cost about $50 to $75 for a small house.

Insulate your attic even if you live in a warm climate. This insulation will help to keep you cool and comfortable during the summer even as it helps to keep you warm in the winter. For a trailer or home where there isn't enough crawl space above the ceiling to install insulation, follow the directions below. In colder climates, you will need more than 3 inches of insulation in the attic, perhaps as much as 6 to 12 inches. The first 3 inches will pay for itself in heat savings very quickly. Use this money to buy even more insulation. Put it on top of the first 3 inches until you have enough.

Ask your Community Action Agency or County Extension Service for advice on how much insulation you should have, and how to install it. The Bibliography at the end of this booklet lists other sources of information. Now, let's look at how to install attic insulation.

ATTIC

First, look into your attic and find out where it's best to insulate. If the attic is unfinished, the insulation should be placed between the ceiling joists. If you have a wooden floor in your attic but don't heat this area, try to remove enough of the floor boards to get the insulation underneath all the floor boards. If taking up these boards is too big a job, you can staple the insulation in place on the underside of the roof, between the rafters. If there are rooms that are used and heated in your attic, put the insulation above the ceiling and behind walls which face out toward unheated areas. If

there is not enough crawl
space to get into the attic
and insulate, or if you
live in a trailer with a
flat ceiling, use "acous-
tical ceiling tiles." These
will help insulate a little
and should not cost much if
you buy them at a surplus
sale. You can glue or
staple these tiles in
place. Glue works better where staples won't hold, as in plaster.

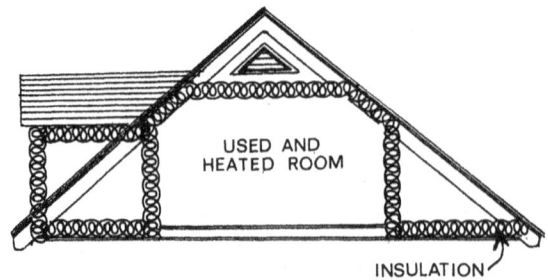

USED AND
HEATED ROOM

INSULATION

If you are unable to do the insulation work yourself, talk
with several contractors and get estimates. Shop carefully and
make sure you get good work done for the money you pay. If you
have any questions about the need for certain work, the quality of
certain materials, or the cost of the work, check with your local
Community Action Agency, County Extension Service, or Better Business Bureau.

If you do the work yourself, there are several types of insulation which you can use in your attic. You can buy insulation in
packages with separate pieces called "batts", in a roll called a
"blanket", or in bags called "loose fill." These materials have
different names. Mineral wool insulation (sometimes called "glass
fiber" or "rock wool") comes as batts, blankets, or loose fill. Cellulose fiber, perlite, or vermiculite insulations come as loose fill.
When you buy insulation, check the "resistance" of the material. Shop
around and buy the insulation which has the greatest resistance per
inch of thickness at the lowest cost. Then buy enough of this material to insulate your whole attic. If you buy blankets or batts, try
to get the kind which has a "vapor barrier."

Once you have the insulation, prepare for the job. Wear loose-
fitting old clothes, with a long-sleeved shirt and gloves. Before you
go up into the attic, get the right tools. You'll need a sharp knife
with a serrated edge to cut batts and blankets. Bring a measuring
tape and a rake. The rake will help to push batts and blankets into
places where there isn't much headroom, or to smooth out loose fill
insulation. A portable light with an extension cord will be handy in
dark areas of the attic. If you must insulate the underside of the
roof, borrow a staple gun. Also, if you don't have an attic floor,
get some plywood boards to walk and kneel on, so you don't go through
the ceiling. These should be at least 1/2 inch thick, and about 2
feet wide by 4 feet long. When you get up into the attic to start
work, watch out for nails coming through the roof over your head.
Don't smoke while in the attic!

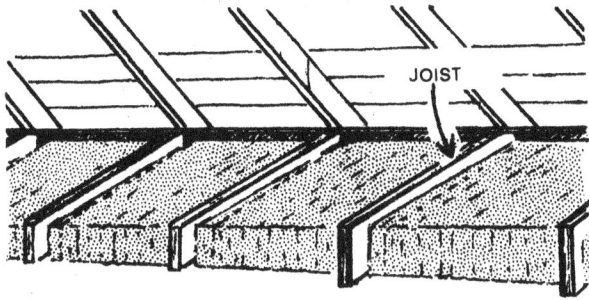

Now, if your attic doesn't have a floor, place batts or blankets, or pour loose fill insulation, between the joists. Be careful not to cover the eaves vents. If your batt or blanket insulation has a vapor barrier, the vapor barrier must face down. For loose fill insulation, or batts and blankets without a vapor barrier, cut up clear "polyethylene" plastic into strips that fit between the joists. Put the plastic in place before you lay down the insulation. Also, some batts come with paper on the side that does not have the vapor barrier. This helps protect the insulation during shipment. Strip the paper off after the batt is in place.

Place insulation on top of the attic hatch cover. Staple it in place around the edge. Insulate under attic walkways, too. Remove sideboards from the walkway if you have to, and push the insulation into place. If you are adding a second layer of insulation to the attic, and use a batt or blanket with a vapor barrier, slash the vapor barrier with a knife several times. This will allow the insulation to dry out if moisture builds up.

While you are in the attic, make sure to insulate hot water pipes and hot air ducts with wraparound insulation. On ducts, check for air leaks and seal them with duct tape before you insulate.

Fill the space between a chimney and any wood framing with materials that can't burn. A mineral wool batt or blanket (without any covering paper) would be good. Also, keep the insulation at least 3 inches away from recessed light fixtures or other heat-producing equipment. The heat buildup might damage the equipment, and it could even start a fire if the insulation is the type that burns.

Make sure your insulation stays dry. A vapor barrier will help. Even if you use a vapor barrier when you put the insulation in, check your insulation at the end of one winter. If there is moisture buildup, open your attic windows an inch or so. Another way to keep your attic dry is to use vents in the eaves or ridge area. In a small house, use about 2 square feet of vents. Put several vents in different places along the eaves or ridges so that air can keep all parts of the attic dry. These vents will also help to keep your house cool in summer. They stop hot air from building up in the attic.

WALLS AND FLOORS

Once you have enough insulation in your attic, insulate behind your walls and under your floors, if the basement or crawl space is unheated. If you have the money to buy materials or hire a contractor to do the work, this insulation is a good investment. Ask your local Community Action Agency or County Extension Service for advice about how much insulation you need and what type to use. Also, check with them or the Better Business Bureau if you have any questions about a contractor.

You can place insulation under your floor yourself, using chicken wire to hold it up. On walls, you can insulate with batts or blankets when you renovate. If a contractor does this work, he may use a machine to blow in a mineral wool or cellulose type of insulation through holes drilled into the wall. Or he may use a plastic foam type of insulation which is installed by a machine under pressure.

If you don't have the money for these insulation materials or can't afford to hire a contractor, you can still help yourself. Make sure cold air doesn't blow up through cracks in the floor. Put skirts on trailers, or bank around an open foundation. Rugs on the floor will help stop cold air leaks and will also help insulate. See page 8 for more information.

Hot water heating, stoves, and refrigerators use up $1 out of every $5 that you spend for energy. Let's see if you can save some dollars here as well.

HOT WATER HEATERS

Don't overheat your water. Use the lowest temperature you can. Most people find that 110° or 120° is good. Turn down the thermostat on your water heater. There is a dial for this, usually near the pilot light. Turn it from a "High" or "Hot" setting to a "Low" or "Warm" setting. On electric hot water heaters, there may be two heating elements. Turn down the thermostat on both. Also, drain about a gallon of water from the bottom of the tank once each year. This will remove sediment that can insulate the water from the heating element.

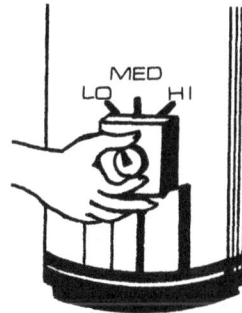

INSULATE HOT WATER PIPE

Insulate your hot water tank. Just a few dollars of blanket insulation can save you $10 or more each year. Get blanket insulation with a paper backing. Wrap it around the tank with the paper facing out, and tape the joints to hold the insulation in place. Leave enough space around the drain plug, temperature controls, and the face-plate which may cover a pilot light, so you can get at these. Also, on oil and gas heaters, make sure there is space so air can get into the burner area. Caution! Do not put insulation on the top of a gas-fired water heater! The heat near the flue may cause the paper backing on the insulation to burn.

Insulate the pipe between the hot water heater and the faucet. This helps keep the hot water in the pipe warm. Either pipe insulation, or wrap-around insulation, is good for this. You can buy either at most hardware stores.

Use as little hot water as you can for washing. For dishes, fill the basin or a small tub with hot soapy water once. Then wash. Rinse dishes with cold water. The same is true for washing hands and face, and shaving. Fill the basin once. Don't let hot water run. After washing dishes or bathing during the winter, leave the hot water in the sink or tub until it cools off. It will help heat the room, and put a little moisture into the air.

Fix leaky faucets. These waste both hot and cold water. They cost you money. Usually a drip can be stopped by replacing a washer.

REFRIGERATORS

These are next in line for attention. Here are some things you can do to save dollars:

Turn the refrigerator to the warmest setting that will keep food from spoiling. There is a dial inside the refrigerator for this, and you can check the temperature with a thermometer. 40° is fine for the refrigerator, and 10° for the freezer.

GASKET

Make sure the gasket on the refrigerator door closes tightly. To test for this, close the door on some newspaper. If the paper pulls out easily you need a new gasket. Also, make sure that the door on the freezer closes correctly.

Open and close the refrigerator doors as little as possible. Don't let children play here.

Most foods should be allowed to cool down after cooking outside the refrigerator. A few recipes call for rapid chilling, but most things can be cooled in the house. This reduces the load on the refrigerator.

If the refrigerator is near a heater or a sunny window, move it to a cooler spot in the kitchen.

COOKING

Don't use the burners or the oven to heat the kitchen. This wastes a lot of heat because the stove is less efficient than a furnace. It's also dangerous! A draft could blow out the oven pilot light on a gas stove. And on both gas and electric stoves, too high a temperature for long periods of time could ignite grease in the oven. Never leave the kitchen stove on when you are out of the kitchen for long periods or asleep! If you need extra heat, or in case of emergency, the small "chunk stoves" mentioned earlier or an electric blanket will do a lot better job.

If possible, turn off the burner pilot lights on a gas stove. These eat up gas all the time, whether you are using the burners or not. A fourth of your cooking bill can be wasted on the pilot lights. You can light the burners as you use them with kitchen matches or, better yet, a flint burner lighter. <u>Never just blow the pilot light out!</u> Be sure the gas to the pilot light is "off", by shutting off its valve on the burner with a screwdriver.

In the winter, roast food slowly in the oven at a low temperature (275° or below). This saves cooking fuel, preserves nutrition, and helps heat the house. In summer, foods should be boiled or sauteed quickly on the burners whenever possible, to keep the kitchen cool. Foods that must be cooked a long time, and can be cooked either on top of the stove or in the oven, should be oven-cooked. Also, try to cook several foods at once in the oven. Choose foods that cook at nearly the same temperature. You can cook two or three dishes for little more than the cost of one this way.

When you use the burners, choose pans that have wide, flat bottoms that just cover the burner and absorb all the heat. When water is used in cooking, such as for vegetables, use just enough water to steam them and prevent sticking -- and cover the pot with a lid. Don't overcook. This wastes energy and also food value.

Although pressure cookers cost more to begin with, they cook food quicker than ordinary pots and use much less gas or electricity. The food will have a nicer appearance and flavor. Cheaper cuts of meat can be made tender by pressure cooking.

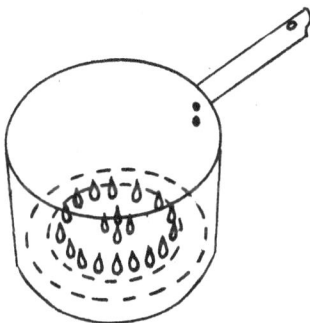
POT TOO BIG

BAD

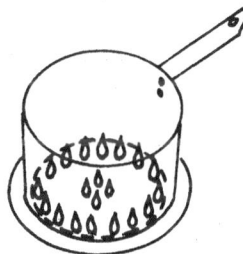
POT JUST RIGHT

GOOD

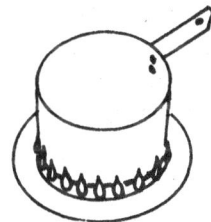
POT TOO SMALL

BAD

<u>Remember</u>: Put the ideas in this booklet to work for you. You can save as much as $60 or more out of each $100 you now spend for heat. It's worth the effort!

DO-IT-YOURSELF PROJECTS

Once you have put the other ideas in this booklet to work, you may want to try some new projects. They are fun to build, low in cost, and can provide you with even more heat and comfort.

FIREPLACES

Many folks use fireplaces for extra heat. Usually a lot of wood has to be burned because fireplaces are not very efficient. This happens for two reasons. First, even small fires draw large amounts of cold air into the house to feed the fire. This creates cold drafts along the floor and cools the house. Second, as soon as the heat comes off the fire, it is drawn up the flue. Not much heat gets out into the room.

Because of these problems, you have to cut or purchase a lot of firewood. You won't get very much heat for all your work. The two low-cost projects outlined here, however, will help. If you put them both to work for you, it's possible to almost double the heat you get from your fireplace! (Of course, no matter how well you fix up your fireplace, you would still get greater efficiency from a wood heating stove, as described earlier in this booklet.)

FIREPLACE AIR VENT

Put a vent under your floor to feed air to the fireplace (or wood stove). This will reduce the amount of cold air which is normally drawn across the room from cracks around windows and doors. This is particularly important after you have weatherstripped your windows and doors, and added storm windows. It will now be harder to pull fresh air into the house for the fire. The fireplace air vent will help to maintain enough of a draft so that smoke won't back up into your home.

The easiest way to connect a vent to the outside is to use a series of 6" circular metal ducts, or their rectangular equivalent. You can buy these ducts at a plumbing supply store. You will need enough to stretch under your floor from just in front of the fireplace to the nearest opening in the basement.

Now, cut a hole for one end of the duct system right in front of the fireplace. Then make an opening in the basement wall or through a basement window for another duct to go through to the outside. When you put the basement duct in place, tilt it down a little toward the out-of-doors. This will allow any moisture to drain out. Also, caulk to make an airtight seal on the outside wall where the duct comes through.

Two elbows are attached to the duct system. One should be placed face down with a screen covering it, just outside the house where the basement duct comes through. This will keep the rain, wind, leaves, and insects out of the duct. The other is used where the fireplace duct comes up to the opening in the floor just in front of the fireplace.

Once you have the ducts in place, you finish the job by placing a grille over the opening in the floor. Use a strong grille so you can walk on it, and get one which can be closed. When the fireplace is not in use, close the grille and put a throw rug over it.

FIREPLACE AIR VENT   FROM ABOVE

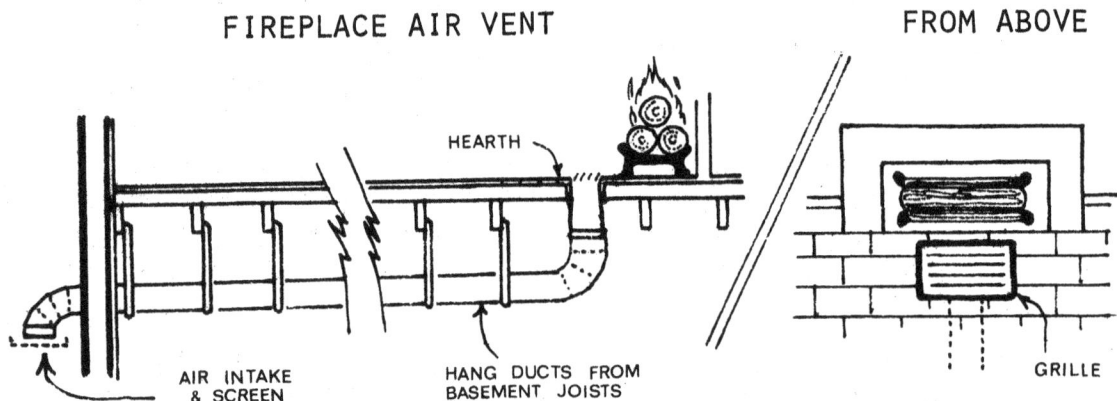

HEARTH

AIR INTAKE
& SCREEN

HANG DUCTS FROM
BASEMENT JOISTS

GRILLE

AIR HEATING GRATE

The next step is to capture more of the fire's heat before it goes up the flue. There is an easy way to do this. You can build (or perhaps buy at a pre-season sale) an air heating grate which not only holds the burning wood but also forces warm air out into the room. This type of grate will cool down the flue gases and might cause your fireplace to smoke and soot up. Be sure your fireplace normally has a good draft before you use it, and make certain that your chimney or fireplace is kept clean, as explained earlier in this booklet.

The grate is made of 1-inch black iron pipes bent into a sort of "C"

shape. The vertical pieces
are bent around so that the
top of the pipes extends just
outside the fireplace. Pack
sand in these vertical pipes
when you bend them, so they
won't crimp.

Pipe elbows can be used
to make the connection between
these vertical pieces and the
bottom pipe sections of the
grate. Make the bottom pieces
just long enough to come out
in front of the mantel a few
inches, but stay behind your
new air vent grille.

Make enough of these "C"
shaped pipes to come all the
way across the face of your
fireplace, with a 3-inch to
4-inch space between each "C". The "C's" rest on and are attached to a
base made of black iron pipe also. The legs of the base should be 2 inches
to 3 inches high and can be made by bending the ends of the pipes. Have the
"C" shaped pipes welded to the base at a local welding shop. The cost should
be only a few dollars.

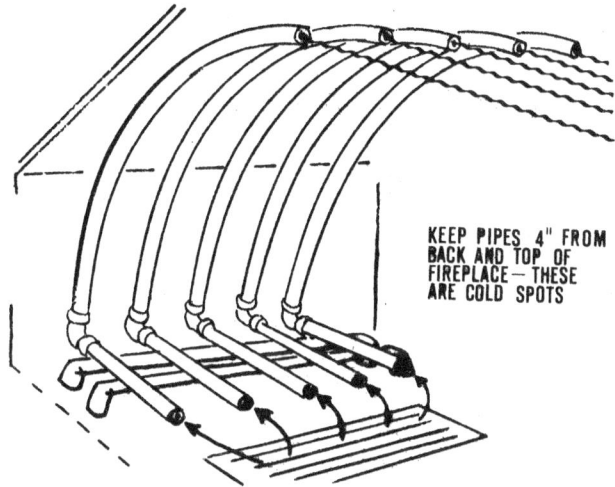

KEEP PIPES 4" FROM
BACK AND TOP OF
FIREPLACE — THESE
ARE COLD SPOTS

AIR HEATING GRATE

This air heating grate works by sucking cold air from the floor. The
air is heated by the fire, and then flows out into the room.

ADD A SMALL PORCH

Building a small porch around the door on the south side of your house
can really be fun. It has many uses, too. The little porch saves heat. It
acts as an air lock in the winter, keeping the wind from blowing cold air
into the house each time you open the door. With windows built into the
three other sides of the porch, it's nice and warm when the sun shines -- a
good place to dry out wet shoes and overcoats. In the winter, it's a good
storage space for firewood. Come spring, the warmth from the sun through
the windows makes it a great little greenhouse.

First, check the local building codes. Building this porch should be
easy and inexpensive. Keep your porch small. It's small enough that you
can use scrap materials and it doesn't need to be insulated.

Start by making two foundation footings of stone or brick.
They should stick up about 4" above the ground and go deep enough
to make a strong support. Now build the porch floor about 6' by 6'

with some old beams and plywood. Add skirts to cover the space between the floor and the ground. These will keep cold air from blowing up under the floor. The walls can be made with scrap two-by-fours, some old siding and storm windows. These and other materials for the roof can be purchased very cheaply at a garage sale or auction. The porch door should open out to make a good exit in case of fire.

The easiest roof to work with is one that is pitched. Make it steep enough so it doesn't collect too much snow. When you build the porch, don't worry about insulation unless you want to use it as a workroom in the winter. It's important, though, that it be tightly constructed so the wind can't blow in. Caulking around all the seams will be a big help.

AIR LOCK PORCH

If you don't need the greenhouse or storage space, you could use this same method of construction to make a smaller 3' by 3' airlock.

MORE HEAT FROM THE SUN

We have already described how to use the sunny windows on the south side of your house for heating. Now you can get even more of the sun's heat by building special solar heaters into your house.

These heaters draw cool house air from the floor into a place where it is heated up by the sun, then push the warmed air back into the house. There are two ways to build them. One kind can be added to your windows, and the other kind makes use of the walls between the windows. Even if your house faces a little to the east or west of true south, they will still work. If you build a solar heater, you will be surprised at how much extra heat you can get -- almost enough to heat one whole room of your house on sunny winter days.

THE WINDOW HEATER

This heater is made of a wooden box with a glass top. It fits into your window. The box is divided into an upper and a lower section by a plywood sheet which is painted black on top. As the sun comes through the glass, the air over the black plywood is heated and flows into the house. This pulls cooler house air from the floor into the heater, where it also is heated.

To build the heater, first build a shell just wide enough to make a tight fit inside your south window. You can use 1" by 10" lumber for the sides and 3/8" exterior plywood for the bottom. Make sure all seams are air-tight and will keep out moisture.

Now add the black plywood divider to this box. Leave a 4" space at the foot of the heater so the air can get around to be heated. To make this divider, use a sheet of 3/8" exterior plywood. Measure it to fit across the inside of the shell. Nail a backing to the divider made of exterior insulation board at least 1/2" thick. Paint the top of the plywood with flat black paint. Now build some supports with furring strips so that the divider will lie about 3" below the glass. Nail the divider into place.

With this divider in the box, you are almost finished. Next, add a piece of 1" x 10" wood for the window to rest on when closed. This piece stretches over the whole width of the box and is nailed to the top of the 1" x 10" sides of the shell. Make a solid vent flap in this top piece just outside the window, using a piano hinge. This vent will be opened in summer to help cool your house. Now, add a small plywood "apron" where the bottom section comes into the house. This makes sure that only cool air from the floor gets pulled into the heater. To do this, stretch a piece of plywood across the mouth of the heater so that it seals tightly against the divider. Then make a flap door so that the upper section of the heater which comes into the house can be closed off. Use plywood and a piano hinge.

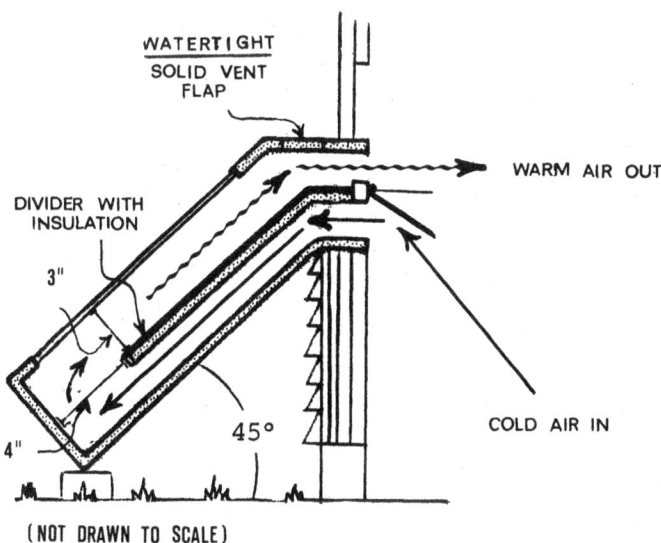

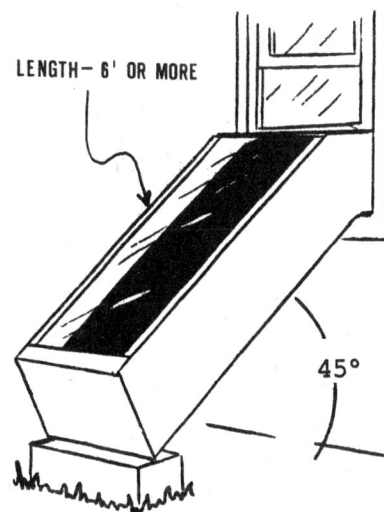

WATERTIGHT SOLID VENT FLAP

DIVIDER WITH INSULATION

3"

4"

45°

WARM AIR OUT

COLD AIR IN

(NOT DRAWN TO SCALE)

LENGTH- 6' OR MORE

45°

WINDOW HEATER

The next step is to add the glass cover to the heater. It may be easier and cheaper to use two or three pieces of glass to cover the heater, rather than one larger sheet. If you do, you must nail one or two wood supports across the heater's width, for the glass to rest on. Then have the glass cut to fit. Place it on and hold it in place with glazier points. Don't putty it. Instead, use duct tape stapled to the sides of the box, or caulk the glass all the way around the edge so you get a really tight seal.

Just one more step: paint the wood with house paint before you put the heater in your window. When you use the heater, be sure to get an airtight fit between the window frame and the heater, and at the top of the lower sash. Use cloth or weatherstripping to do this.

In winter, close off the top flap door at night. During the summer, you can store this window heater in your garage or basement. Or, close the top flap inside the house and open the outside vent. This will help to ventilate the house. When doing this, be sure to open a window across the room for cross-ventilation. That window should be on the shady side so that the heater will draw cooler outside air into the house.

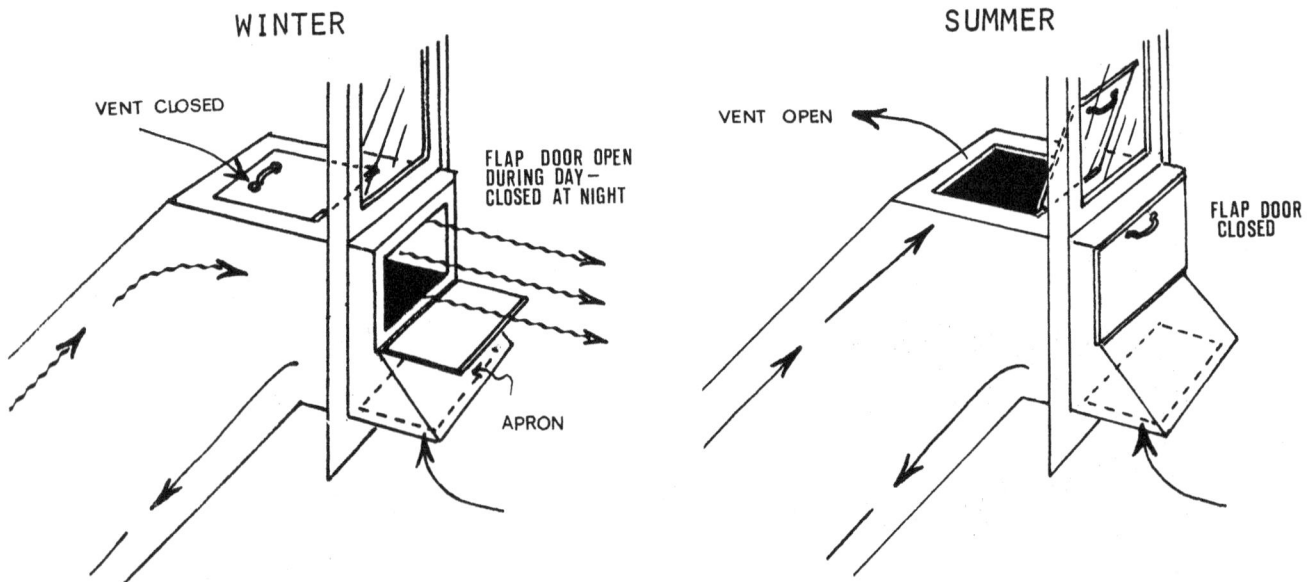

WINTER

VENT CLOSED

FLAP DOOR OPEN
DURING DAY —
CLOSED AT NIGHT

APRON

SUMMER

VENT OPEN

FLAP DOOR
CLOSED

THE WALL HEATER

This type of solar heater makes use of the wall space on the south side of your house. It may cost less to build than the other heater if you make use of old storm windows. However, you must cut some holes in your wall to let cold air in at the bottom and warm air out at the top.

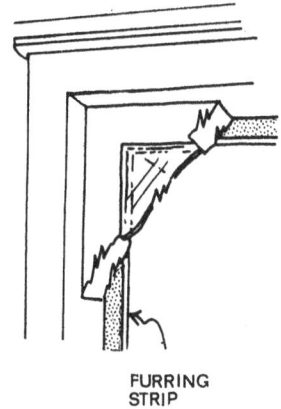

To build one of these heaters, first try to get some old storm windows. This way the glass won't cost much. Once you have the windows, keep them in their frames. Plan the size of the wall heater so that the window frames will just cover the front of the heater.

FURRING STRIP

BUILD HEATER OUT OF OLD STORM WINDOWS

Now, clean the part of the house wall which the heater will cover. Screw 2" x 4"'s flat on the wall to make a box around this space. If you have a brick or masonry wall, attach the 2" x 4"'s with masonry anchors. Inside the 2" x 4"'s, make vent holes at the top and bottom of the heater for air to come in and out. If your heater is about 3 feet wide, a hole at the top and another at the bottom, about 4" deep by 14" wide, will work fine. If it is 6 feet wide, you will need two sets of holes, spaced about 3 feet apart. Once this is done, paint the area inside the 2" x 4"'s with flat black paint.

Next, build a box butting up against the outside of the 2" x 4"'s. Use 1" x 6" lumber. Seal at the edge where it meets the house with caulking, to close any cracks. Put some concrete blocks or bricks beneath the heater to help hold up its weight. Now, nail pieces of furring strip to the inside of the 1" x 6" box. Place these several inches from the outer edge. They serve as back supports for the storm windows.

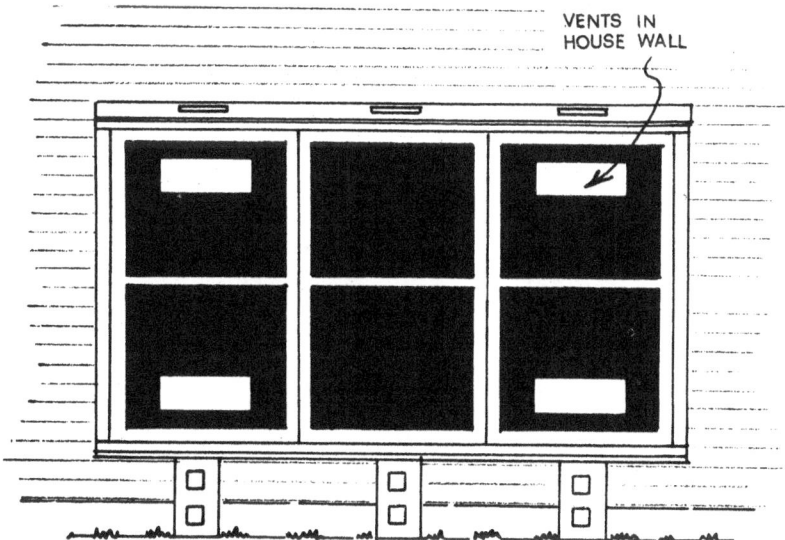

FRONT VIEW   WALL HEATER

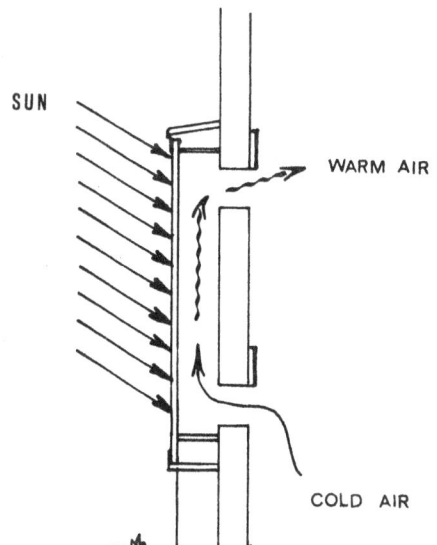

SIDE VIEW

Then put the storm windows against the furring strips on the heater box. Screw them in place from the outside. If you can get enough of these, make double panes for more efficiency. For support, place 2" x 4" spacers behind where the storm windows meet, and screw them in place. Caulk all around the edges. Paint the outside with house paint -- and you're finished!

You will want to close the hole on the bottom at night so warm house air doesn't cool off. Just stuff some cloth into it. When the snows come, shovel snow around the front of the wall heater. It will reflect sunlight into it. A reflector made with old mirrors or printing plates stapled to a plywood sheet will help even more. During the summer, you can cover this wall heater with a bamboo screen. Or if you put solid vent flaps in the top of the box, you can open these (while closing off the upper vent holes inside the house) and use your heater to ventilate the house during the summer. When doing this, use another window on the shady side of the house for cross-ventilation.

EMERGENCIES !

If your heater breaks down or you run out of fuel, there are several things you should do:

* Try to make simple repairs if your heater is broken. Locate the maintenance booklet or read the instructions from the faceplate on the unit. If you can't get your heater to operate by following these instructions, call in a repairman.

* If you're out of fuel, call the fuel dealer and ask for an emergency delivery. If he can't bring it for some reason, the information on the inside back cover of this booklet will tell you where to call for more help.

* If you have an oil furnace and keep an emergency supply of oil in a 5-gallon can, carefully pour this into the fuel-fill line that leads to the tank. Re-start the furnace if necessary, and turn down the heat to 50°. Five gallons should last a day or more till you can get a delivery. Some people use a gas or electric stove in an emergency to keep the kitchen warm. Be careful, as this can be dangerous. A draft could blow out the pilot light in an open gas-oven door. And on both gas and electric stoves too high a temperature for long periods of time could ignite grease in the oven. Never leave the kitchen stove on when you are out of the kitchen or asleep!

If you just can't get a repairman or fuel for several hours or days, take these steps:

* Get dressed in very warm clothes.

* Once you are warm, heat only the rooms which are absolutely needed.

* Use your fireplace or wood heater, or wrap up in an electric blanket, if you have one. If not, try to borrow a small space heater, camp stove, or charcoal grill. Remember, if these are not electric, you must have plenty of air ventilation when you use them!

* Fill pots and pans with water for drinking. Then drain your water pipes. Shut off the valve where the water comes into the house. Open the faucets and close the water inlet valve on the toilet. Now, find the drain valve for the water system and let the water out. It's usually in the basement. Then, flush the toilet and put some crankcase oil or kerosene in the bowl. The weight of these will flush out the trap section and prevent freezing. If you have hot water radiators, make sure they drain by letting in air at the valve. Taking these steps may prevent the water in pipes and fixtures from freezing and causing damage.

* If you need further help right away, ask your neighbors.

Each fall, before winter starts, make sure your home is ready. Read this list and do any work that is necessary.

* Are you ready for an emergency if your heat goes off and you can't get fuel? Do you have extra firewood or oil if you use these for heat? Do you know where to call for help and do you know how to drain water pipes to prevent freezing? Read page 45. Make certain you are ready for emergencies!

* Will your weatherstripping and caulking keep out the cold wind this winter? Check around windows and doors on a windy day, and replace weatherstripping or caulking where necessary. Also, have you put up your storm windows? See pages 5 to 9 for information on how to do this work.

* If you plan to close off extra doors, windows, or rooms for the winter, have you done this yet? See pages 13 to 15.

* Have you put up curtains to keep heat away from the glass at night? See pages 19 to 21.

* Is your heating system ready for winter? If you use wood or coal for heat, are the flue and chimney clean? Are radiators and hot air outlets clean so heat can get out into the room? See pages 23 to 25.

* Do you have enough insulation in your attic? Add insulation now. This way you can save money all winter. See pages 29 to 31.

* Have you drained the sediment from the bottom of your hot water heater this year? Have you checked the gasket on your refrigerator door? See pages 33 and 34.

* Did you have any problems last winter that you couldn't handle? Ask your neighbors for advice and help. Or, call your local Community Action Agency or County Extension Service. They may be able to answer questions or give you assistance.

You will get some good ideas in these magazines, brochures and booklets:

Magazines like Popular Mechanics, Popular Science, Mechanix Illustrated, and Family Handyman often have articles on things people can do to save energy. You can look through these magazines at your library. While there, you could also ask your librarian for the latest books on home energy conservation.

Contact the local Extension Service of your state University. They may have brochures and booklets with helpful information. You can get booklets like these:

How to Set Up a Coal or Wood Stove, a 2-page brochure;

Home Heating in an Emergency, a 15-page booklet.

> Single copies of both are
> available free from the:
> Cooperative Extension Service
> University of New Hampshire
> Durham, NH 03824

Making the Most of Your Energy Dollars in Home Heating and Cooling, a 16-page booklet with information about insulation and the cost savings possible with energy conservation in the home. Single copies are available free from:

> Publications Office
> National Bureau of Standards
> Washington, DC 20234

A Community Planning Guide to Weatherization, a 37-page booklet that explains how to set up a cost-effective weatherization program in your neighborhood. Single copies are available free from:

> Community Services Administration
> Washington, DC 20506

The Fuel Savers: A Kit of Solar Ideas for Existing Homes, a 60-page booklet with ideas and drawings for do-it-yourself solar heating projects. Copies are available for $3.50 from:

> NORWESCAP, Inc.
> Prospect St.
> Phillipsburg, NJ 08865

CSA REGIONAL OFFICES

(Ask for Regional Energy Coordinator)

BOSTON REGIONAL OFFICE - I
John F. Kennedy Federal Building
Room E-400
Boston, Massachusetts 02203
(617) 223-4019
States served: Maine, New Hampshire,
Vermont, Massachusetts, Connecticut,
and Rhode Island

NEW YORK REGIONAL OFFICE - II
26 Federal Plaza, 32nd floor
Room 3227
New York, New York 10007
(212) 264-3960
States served: New York, New Jersey,
Puerto Rico, Virgin Islands

PHILADELPHIA REGIONAL OFFICE - III
Gateway Building, Room 2400
3535 Market Street
Philadelphia, Pennsylvania 19104
(215) 596-6022
States served: Pennsylvania, Maryland,
Delaware, Virginia, West Virginia,
District of Columbia

ATLANTA REGIONAL OFFICE - IV
730 Peachtree Street N.E.
Atlanta, Georgia 30308
(404) 881-3526
States served: Kentucky, Tennessee,
North Carolina, South Carolina,
Georgia, Alabama, Mississippi,
Florida

CHICAGO REGIONAL OFFICE - V
300 South Wacker Drive, 26th floor
Chicago, Illinois 60606
(312) 353-5988
States served: Illinois, Wisconsin,
Minnesota, Michigan, Indiana, Ohio

DALLAS REGIONAL OFFICE - VI
1200 Main Tower
Dallas, Texas 75202
(214) 749-1381
States served: Texas, Oklahoma,
Louisiana, Arkansas, New Mexico

KANSAS CITY REGIONAL OFFICE - VII
911 Walnut Street, Room 1600
Kansas City, Missouri 64106
(816) 374-3561
States served: Missouri, Kansas,
Iowa, Nebraska

DENVER REGIONAL OFFICE - VIII
Federal Building 12th floor
1961 Stout Street
Denver, Colorado 80294
(303) 837-4923
States served: Colorado, Utah, Wyoming,
Montana, North Dakota, South Dakota

SAN FRANCISCO REGIONAL OFFICE - IX
450 Golden Gate Avenue
Box 36008
San Francisco, California 94102
(415) 556-5400
States served: California, Nevada,
Arizona, Hawaii

SEATTLE REGIONAL OFFICE - X
Arcade Plaza Building
1321 Second Avenue
Seattle, Washington 98101
(206) 442-0183
States served: Washington, Oregon,
Idaho, Alaska

FEDERAL ENERGY ADMINISTRATION
Office of Consumer Affairs/Special Impact
12th and Pennsylvania Avenue N.W.
Room 4310
Washington, DC 20461
(202) 566-9021

www.ingramcontent.com/pod-product-compliance
Lightning Source LLC
Chambersburg PA
CBHW081303040426
42452CB00014B/2631